Binary, Octal and Hexadecimal

for Programming & Computer Science

Copyright & Other Notices

Introduction

This book introduces the binary, octal and hexadecimal numbering systems used in computer science and computer programming. It introduces how numbers are represented in each of these systems, how to convert between them (and to and from base 10).

After completing this book, you might wish to look at my follow-up book, Advanced Binary for Programming & Computer Science: Logical, Bitwise and Arithmetic Operations, and Data Encoding and Representation, which introduces some more advanced concepts related to binary, including various logical, mathematical and arithmetic operations that are performed on binary numbers, and how binary can be used to store characters, text, negative and real numbers including fixed-point and floating-point representations, and binary coded decimals.

When I began writing this book, I had already written several other books on various math topics and did consider whether this book should be considered a math book or a computing book. After some thought, I decided on the latter, since the mathematical aspects of these number bases would have to be covered in any case, but I did not wish to ignore the additional considerations that computer programmers and scientists need to consider.

In this book, among other things, you will learn:

- What are number bases (also known as radixes)

- What is binary (base 2)

- How to convert binary numbers to denary (base 10)

- How to convert denary numbers to binary

- What is octal (base 8)

- How to convert octal numbers to denary

- How to convert denary numbers to octal

- Why many programmers and computer scientists use octal

- How to convert octal numbers to binary

- How to convert binary numbers to octal

- What is hexadecimal (base 16)

- How to convert hexadecimal numbers to denary

- How to convert denary numbers to hexadecimal

- Why many programmers and computer scientists use hexadecimal

- How to convert hexadecimal numbers to binary

- How to convert binary numbers to hexadecimal

- Is there a reason to prefer octal over hexadecimal or vice-versa?

As I said earlier, after completing this book, I would suggest my follow-up book, Advanced Binary for Programming & Computer Science: Logical, Bitwise and Arithmetic Operations, and Data Encoding and Representation, if you wish to study more advanced topics.

I hope you enjoy this book

- If you want to find out about my math books, please visit: http://www.suniltanna.com/math

- For my science books, please visit: http://www.suniltanna.com/science

- For my computing books, please visit: http://www.suniltanna.com/computing

Chapter 1: Introducing Number Bases

When we think about numbers, most of the time we rarely consider how they are represented and whether there might be an alternative way to represent the same number.

Let us consider the number 27056:

What do the individual digits mean?

Ten Thousands $(\times 10^4)$ $\times 10000$	Thousands $(\times 10^3)$ $\times 1000$	Hundreds $(\times 10^2)$ $\times 100$	Tens $(\times 10^1)$ $\times 10$	Units $(\times 10^0)$ $\times 1$
2	7	0	5	6

- The 6 in the units column represents 6 because $6 \times 10^0 = 6 \times 1 = 6$.
- The 5 in the tens column represents 50 because $5 \times 10^1 = 5 \times 10 = 50$.
- The 0 in the hundreds column represents 0 because $0 \times 10^2 = 0 \times 100 = 0$.
- The 7 in the thousands column represents 7000 because $7 \times 10^3 = 7 \times 1000 = 7000$.
- The 2 in the ten thousands column represents 20000 because $2 \times 10^4 = 2 \times 10000 = 20000$.
- The total value of the number can be calculated by added the column values together: 6 + 50 + 0 + 7000 + 20000 = 27056.

Now consider the number 138:

What do the individual digits mean?

Ten Thousands $(\times 10^4)$ $\times 10000$	Thousands $(\times 10^3)$ $\times 1000$	Hundreds $(\times 10^2)$ $\times 100$	Tens $(\times 10^1)$ $\times 10$	Units $(\times 10^0)$ $\times 1$
		1	3	8

- The 8 in the units column represents 8 because $8 \times 10^0 = 8 \times 1 = 8$.
- The 3 in the tens column represents 30 because $3 \times 10^1 = 3 \times 10 = 30$.
- The 1 in the hundreds column represents 100 because $1 \times 10^2 = 1 \times 100 = 100$.
- The total value of the number can be calculated by added the column values together: 8 + 30 + 100 = 138.

Now consider the number 3947:

What do the individual digits mean?

Ten Thousands (× 10^4) × 10000	Thousands (× 10^3) × 1000	Hundreds (× 10^2) × 100	Tens (× 10^1) × 10	Units (×10^0) × 1
	3	9	4	7

- The 7 in the units column represents 7 because $7 \times 10^0 = 7 \times 1 = 7$.
- The 4 in the tens column represents 40 because $4 \times 10^1 = 4 \times 10 = 40$.
- The 9 in the hundreds column represents 900 because $9 \times 10^2 = 9 \times 100 = 900$.
- The 3 in the thousands column represents 3000 because $3 \times 10^3 = 3 \times 1000 = 3000$.
- The total value of the number can be calculated by added the column values together: 7 + 40 + 900 + 3000 = 3947.

I could go on, giving you more and more examples, but I hope that there are three things that you have noticed by now:

- The actual value represented by each digit in the number depends on the combination of which digit it is, and on that digit's position within the number. This is known as place value.
- Each successive column represents numbers 10 times bigger than the last. In other words, the columns correspond to the powers of 10.
- Each column can hold 10 possible digit values - from 0 to 9 inclusive.

An obvious question is do numbers have to be represented by columns based on successive powers of 10, with each column holding one of 10 possible digit values? Or are there are other ways to represent numbers?

The short answer is that there are other ways to represent numbers, and this is what this book is about! These different systems are known as number bases or radixes.

- The familiar number system that you use every day, and which is based on powers of 10 and uses 10 different digits, is known as base 10, radix 10, decimal, or denary.
- The number system used internally by digital computers, which is based on powers of 2 and uses only 2 different digits, is known as base 2, radix 2, or binary.
- A number system, which is based on powers of 8 and uses 8 different digits, is known as base 8, radix 8, or octal.
- A number system, which is based on powers of 16 and uses 16 different digits, is known as base 16, radix 16, or hexadecimal.

- Other number bases and radixes exist too - and although they all generally work on similar principles, do **not** have many practical uses in computing, so will **not** be considered in this book.

Chapter 2: Binary

Internally, digital computers generally use the binary (also known as base 2 or radix 2) numbering system to represent numbers. This is because binary only has two digits, 0 and 1, and these can easily be represented in electronic circuits - for example a circuit being off might represent 0, and a circuit being on might represent 1.

(A note on terminology: Each digit of a binary number is sometimes called a bit, which is short for Binary digIT).

In the binary numbering system, the columns of the number correspond to powers of 2, and there are two possible digit values: 0 and 1. Thus in binary, numbers are represented by a sequence of 1s and 0s.

Consider the binary number 1001:

Sixteens $(\times 2^4)$ $\times 16$	Eights $(\times 2^3)$ $\times 8$	Fours $(\times 2^2)$ $\times 4$	Twos $(\times 2^1)$ $\times 2$	Units $(\times 2^0)$ $\times 1$
	1	0	0	1

- The rightmost digit (1), which is in the units column, represents 1 because $1 \times 2^0 = 1 \times 1 = 1$.
- The second from right digit (0), which is in the twos column, represents 0 because $0 \times 2^1 = 0 \times 2 = 0$.
- The third from right digit (0), which is in the fours column, represents 0 because $0 \times 2^2 = 0 \times 4 = 0$.
- The fourth from right digit (1), which is in the eights column, represents 8 because $1 \times 2^3 = 1 \times 8 = 8$.
- The total value of the number, **in denary**, can be calculated by added the column values together: $1 + 0 + 0 + 8 = 9$.
- To sum up, the binary number 1001 represents the same number as the denary number of 9.

Now consider the binary number 101:

Sixteens $(\times 2^4)$ $\times 16$	Eights $(\times 2^3)$ $\times 8$	Fours $(\times 2^2)$ $\times 4$	Twos $(\times 2^1)$ $\times 2$	Units $(\times 2^0)$ $\times 1$
		1	0	1

- The rightmost digit (1), which is in the units column, represents 1 because $1 \times 2^0 = 1 \times 1 = 1$.
- The second from right digit (0), which is in the twos column, represents 0 because $0 \times 2^1 = 0 \times 2 = 0$.
- The third from right digit (1), which is in the fours column, represents 4 because $1 \times 2^2 = 1 \times 4 = 4$.
- The total value of the number, **in denary**, can be calculated by added the column values together: $1 + 0 + 4 = 5$.
- To sum up, the binary number 101 represents the same number as the denary number of 5.

Consider the binary number 11011:

Sixteens ($\times 2^4$) $\times 16$	Eights ($\times 2^3$) $\times 8$	Fours ($\times 2^2$) $\times 4$	Twos ($\times 2^1$) $\times 2$	Units ($\times 2^0$) $\times 1$
1	1	0	1	1

- The rightmost digit (1), which is in the units column, represents 1 because $1 \times 2^0 = 1 \times 1 = 1$.
- The second from right digit (1), which is in the twos column, represents 2 because $1 \times 2^1 = 1 \times 2 = 2$.
- The third from right digit (0), which is in the fours column, represents 0 because $0 \times 2^2 = 0 \times 4 = 0$.
- The fourth from right digit (1), which is in the eights column, represents 8 because $1 \times 2^3 = 1 \times 8 = 8$.
- The fifth from right digit (1), which is in the sixteens column, represents 16 because $1 \times 2^4 = 1 \times 16 = 16$.
- The total value of the number, **in denary**, can be calculated by added the column values together: $1 + 2 + 0 + 8 + 16 = 27$.
- To sum up, the binary number 11011 represents the same number as the denary number of 27.

Now consider the binary number 10110:

Sixteens ($\times 2^4$) $\times 16$	Eights ($\times 2^3$) $\times 8$	Fours ($\times 2^2$) $\times 4$	Twos ($\times 2^1$) $\times 2$	Units ($\times 2^0$) $\times 1$
1	0	1	1	0

- The rightmost digit (0), which is in the units column, represents 0 because $0 \times 2^0 = 0 \times 1 = 0$.
- The second from right digit (1), which is in the twos column, represents 2 because $1 \times 2^1 = 1 \times 2 = 2$.
- The third from right digit (1), which is in the fours column, represents 4 because $1 \times 2^2 = 1 \times 4 = 4$.

- The fourth from right digit (0), which is in the eights column, represents 0 because $0 \times 2^3 = 0 \times 8 = 0$.
- The fifth from right digit (1), which is in the sixteens column, represents 16 because $0 \times 2^4 = 1 \times 16 = 16$.
- The total value of the number, **in denary**, can be calculated by added the column values together: $0 + 2 + 4 + 0 + 16 = 22$.
- To sum up, the binary number 10110 represents the same number as the denary number of 22.

As you can see, it takes more digits to represent a number in binary than in denary:

- The number 9 in denary (a single digit), corresponded to a four-digit binary number.
- The number 5 in denary (a single digit), corresponded to a three-digit binary number.
- The numbers 22 and 27 in denary (two digits) corresponded to five-digit binary numbers.

Consequently, even relatively small numbers, when represented in binary, can correspond to long strings of 1s and 0s. For example:

- The denary number 81 corresponds to 1010001 in binary.
- The denary number 746 corresponds to 1011101010 in binary.
- The denary number 947 corresponds to 1110110011 in binary.
- The denary number 23516 corresponds to 101101111011100 in binary.
- The denary number 48137 corresponds to 1011110000001001 in binary.

These long strings of 1s and 0s are fine for computers to deal with automatically, but not easy for humans to transcribe or to communicate to each other. As we will see in Chapters 3 and 4, computer scientists and programmers have found way to use other number bases (such as octal and hexadecimal) as a kind of shorthand for binary.

In the rest of this chapter, we will look at how to convert numbers from binary to denary and vice-versa. Before doing so, here is a table of the powers of 2, which correspond to the values of the columns in binary numbers:

n	2^n	n	2^n
0	1	16	65536
1	2	17	131072
2	4	18	262144
3	8	19	524288
4	16	20	1048576
5	32	21	2097152
6	64	22	4194304
7	128	23	8388608
8	256	24	16777216
9	512	25	33554432
10	1024	26	67108864
11	2048	27	134217728
12	4096	28	268435456
13	8192	29	536870912
14	16384	30	1073741824
15	32768	31	2147483648

How to convert a binary number to denary

Here are the steps for converting a binary number to denary:

(1) Start with the rightmost digit of the binary number and multiply it by 2^0 (that is 1).

(2) Now move to the second from right digit of the binary number and multiply it by 2^1 (that is 2). Add this to the value from your previous step and keep a running total.

(3) Now move to the third from right digit of the binary number and multiply it by 2^2 (that is 4). Add this to the running total from your previous step to get a new running total.

(4) Continue repeating this process until you have processed all the digits of the binary number, each time using the next higher power of 2. The total at the end is the denary value of the binary number.

Example: Converting the binary number 101011 to denary.

(1) The rightmost digit is 1. So, $1 \times 2^0 = 1 \times 1 = 1$.

(2) The second from right digit is 1. So, $1 \times 2^1 = 1 \times 2 = 2$. The value from the previous step was 1, the running total is now $1 + 2 = 3$.

(3) The third from right digit is 0. So, $0 \times 2^2 = 0 \times 4 = 0$. The running total is now $3 + 0 = 3$.

(4) The fourth from right digit is 1. So, $1 \times 2^3 = 1 \times 8 = 8$. The running total is now $3 + 8 = 11$.

(5) The fifth from right digit is 0. So, $0 \times 2^4 = 0 \times 16 = 0$. The running total is now $11 + 0 = 11$.

(6) The sixth from right digit is 1. So, $1 \times 2^5 = 1 \times 32 = 32$. The running total is now $11 + 32 = 43$.

(7) The binary number 101011 corresponds to denary 43.

How to convert a denary number to binary

Here is a method for converting a denary number to binary (this method generates a binary number starting with the rightmost or lowest-value digit):

(1) Divide the denary number by 2, giving a whole number quotient and a remainder. The remainder is the first binary digit. The quotient is saved for the next step.

(2) Divide the quotient from the previous step by 2, to give a new whole number quotient and a remainder. The remainder is the next binary digit, and quotient is again saved for the next step.

(3) Continue repeating the process until you are left with a whole number quotient of 0.

Example: Converting the denary number 45 to binary.

(1) We divide 45 by 2. The quotient is 22 and the remainder is 1. Hence the rightmost binary digit is 1.

(2) We divide 22 by 2. The quotient is 11 and the remainder is 0. Hence the second from right binary digit is 0, giving a binary number, so far, of 01.

(3) We divide 11 by 2. The quotient is 5 and the remainder is 1. Hence the third from right binary digit is 1, giving a binary number, so far, of 101.

(4) We divide 5 by 2. The quotient is 2 and the remainder is 1. Hence the fourth from right binary digit number is 1, giving a binary number, so far, of 1101.

(5) We divide 2 by 2. The quotient is 1 and the remainder is 0. Hence the fifth from right binary digit is 0, giving a binary number, so far, of 01101.

(6) We divide 1 by 2. The quotient is 0 and the remainder is 1. Hence the sixth from right binary digit is 1, giving a binary number of 101101. Since the quotient was 0, we know that we have finished the conversion process.

(7) Thus, the denary number of 45 corresponds to the binary number of 101101.

How to convert a denary number to binary (alternative method)

Here is an alternative method for converting a denary number to binary (this method generates a binary number starting with the leftmost or highest-value digit):

(1) Find the largest value of 2^n (where n is an integer) which is less than or equal to the denary number. Subtract this from the denary number and put a 1 in the corresponding column of the binary number.

(2) Now try 2^{n-1}. If it is less than what is left, put a 0 in the corresponding column of the binary number. If it is equal or greater than what is left, subtract this value, and put a 1 in the corresponding column of the binary number.

(3) Continue repeating the process until you have down all the powers of 2 from 2^n to 2^0.

Example: Converting the denary number 45 to binary.

(1) The largest power of 2 which is less than or equal to 45 is 2^5 ($2^5 = 32$). The binary number begins with a 1 (in the 2^5 column). 45 minus 32 leaves 13.

(2) The next largest power of 2 is 2^4 ($2^4 = 16$). Since 16 is greater than the 13 we have left, the next binary digit is 0 (in the 2^4 column), giving the binary, so far, of 10. We still have 13 left.

(3) The next largest power of 2 is 2^3 ($2^3 = 8$). Since 8 is less than the 13 we have left, the next binary digit is 1 (in the 2^3 column), giving a binary, so far, of 101. 13 minus 8 leaves 5.

(4) The next largest power of 2 is 2^2 ($2^2 = 4$). Since 4 is less than the 5 we have left, the next binary digit is 1 (in the 2^2 column), giving a binary, so far, of 1011. 5 minus 4 leaves 1.

(5) The next largest power of 2 is 2^1 ($2^1 = 2$). Since 2 is greater than the 1 we have left, the next binary digit is 0 (in the 2^1 column), giving the binary, so far, of 10110. We still have 1 left.

(6) The last power of 2 is 2^0 ($2^0 = 1$). Since 1 is equal to the 1 we have left, the last binary digit is 1 (in the 2^0 column), giving the completed binary of 101101.

(7) Thus, the denary number of 45 corresponds to the binary number of 101101.

Questions

1. What is the denary equivalent of the binary number 10010?

2. What is the denary equivalent of the binary number 110011?

3. What is the denary equivalent of the binary number 101010?

4. What is the denary equivalent of the binary number 100101?

5. What is the denary equivalent of the binary number 1011?

6. What is the binary equivalent of the denary number 27?

7. What is the binary equivalent of the denary number 31?

8. What is the binary equivalent of the denary number 14?

9. What is the binary equivalent of the denary number 53?

10. What is the binary equivalent of the denary number 19?

Answers to Chapter 2 Questions

1. 18

2. 51

3. 42

4. 37

5. 11

6. 11011

7. 11111

8. 1110

9. 110101

10. 10011

Chapter 3: Octal

Just as denary is based on using powers of 10 and 10 different digits (the digits from 0 to 9 inclusive), and binary is based on using powers of 2 and 2 different digits (the digits 0 and 1), the octal number system (also known as base 8 or radix 8) is based on using powers of 8 and 8 different digits (the digits 0 to 7 inclusive).

Consider the octal number 265:

4096s $(\times 8^4)$ $\times 4096$	512s $(\times 8^3)$ $\times 512$	Sixty Fours $(\times 8^2)$ $\times 64$	Eights $(\times 8^1)$ $\times 8$	Units $(\times 8^0)$ $\times 1$
		2	6	5

- The rightmost digit (5), which is in the units column, represents 5 because $5 \times 8^0 = 5 \times 1 = 5$.
- The second from right digit (6), which is in the eights column, represents 48 because $6 \times 8^1 = 6 \times 8 = 48$.
- The third from right digit (2), which is in the sixty fours column, represents 128 because $2 \times 8^2 = 2 \times 64 = 128$.
- The total value of the number, **in denary**, can be calculated by added the column values together: 5 + 48 + 128 = 181.
- To sum up, the octal number 265 represents the same number as the denary number of 181.

Consider the octal number 43107:

4096s $(\times 8^4)$ $\times 4096$	512s $(\times 8^3)$ $\times 512$	Sixty Fours $(\times 8^2)$ $\times 64$	Eights $(\times 8^1)$ $\times 8$	Units $(\times 8^0)$ $\times 1$
4	3	1	0	7

- The rightmost digit (7), which is in the units column, represents 7 because $5 \times 8^0 = 7 \times 1 = 7$.
- The second from right digit (0), which is in the eights column, represents 0 because $0 \times 8^1 = 0 \times 8 = 0$.
- The third from right digit (1), which is in the sixty fours column, represents 64 because $1 \times 8^2 = 1 \times 64 = 64$.
- The fourth from right digit (3), which is in the five hundred and twelves column, represents 1536 because $3 \times 8^3 = 3 \times 512 = 1536$.
- The fifth from right digit (4), which is in the four thousand and ninety sixes column, represents 16384 because $4 \times 8^4 = 4 \times 4096 = 16384$.
- The total value of the number, **in denary**, can be calculated by added the column values together: 7 + 0 + 64 + 1536 + 16384 = 17991.

- To sum up, the octal number 43107 represents the same number as the denary number of 17991.

In the two examples that we have looked at so far, you will notice that the octal number and its denary equivalent both have the same number of digits. In general, octal numbers tend to be the same length or only slightly longer than their denary equivalents.

In the rest of this chapter, we will look at how to convert numbers from octal to denary and vice-versa, as well as how to convert from octal to binary and vice-versa (you will shortly discover that converting between octal and binary is very easy - and it is for this reason that octal is used by programmers and computer scientists). Before doing so, here is a table of the powers of 8, which correspond to the values of the columns in octal numbers:

n	8^n
0	1
1	8
2	64
3	512
4	4096
5	32768
6	262144
7	2097152
8	16777216
9	134217728
10	1073741824
11	8589934592
12	68719476736
13	549755813888
14	4398046511104
15	35184372088832

How to convert an octal number to denary

Here are the steps for converting an octal number to denary:

(1) Start with the rightmost digit of the octal number and multiply it by 8^0 (that is 1).

(2) Now move to the second from right digit of the octal number and multiply it by 8^1 (that is 8). Add this to the running total from your previous step to get a new running total.

(3) Now move to the third from right digit of the octal number and multiply it by 8^2 (that is 64). Add this to the running total from your previous step to get a new running total.

(4) Continue repeating this process until you have processed all the digits of the octal number, each time using the next higher power of 8. The total at the end is the denary value of the octal number.

Example: Converting the octal number 3047 to denary.

(1) The rightmost digit is 7. So, $7 \times 8^0 = 7 \times 1 = 7$.

(2) The second from right digit is 4. So, $4 \times 8^1 = 4 \times 8 = 32$. The value from the previous step was 7, the running total is now $7 + 32 = 39$.

(3) The third from right digit is 0. So, $0 \times 8^2 = 0 \times 64 = 0$. The running total is now $39 + 0 = 39$.

(4) The fourth from right digit is 3. So, $3 \times 8^3 = 3 \times 512 = 1536$. The running total is now $39 + 1536 = 1575$.

(7) The octal number 3047 corresponds to denary 1575.

How to convert a denary number to octal

Here are the steps for converting a denary number to octal (this method generates an octal number starting with the rightmost or lowest-value digit):

(1) Divide the denary number by 8, giving a whole number quotient and a remainder. The remainder is the first octal digit. The quotient is saved for the next step.

(2) Divide the quotient from the previous step by 8, to give a new whole number quotient and a remainder. The remainder is the next octal digit, and quotient is again saved for the next step.

(3) Continue repeating the process until you are left with a whole number quotient of 0.

Example: Converting the denary number 3533 to octal.

(1) We divide 3533 by 8. The quotient is 441 and the remainder is 5. Hence the rightmost octal digit is 5.

(2) We divide 441 by 8. The quotient is 55 and the remainder is 1. Hence the second from right octal digit is 1, giving an octal number, so far, of 15.

(3) We divide 55 by 8. The quotient is 6 and the remainder is 7. Hence the third from right octal digit is 7, giving an octal number, so far, of 715.

(4) We divide 6 by 6. The quotient is 0 and the remainder is 6. Hence the fourth from right octal digit number is 6, giving an octal number of 6715. Since the quotient was 0, we know that we have finished the conversion process.

(5) Thus, the denary number of 3533 corresponds to the octal number of 6715.

Why programmers and computer scientists use octal

At this point you are probably thinking that digital computers use binary numbers internally and that humans generally prefer denary numbers, so why bother with octal numbers?

The answer is that octal numbers are almost as short and easy-to-read as denary numbers, but unlike denary, octal numbers can **very** easily be converted to and from binary.

The process of converting an octal number to binary, or vice-versa, is in fact so easy that with a bit of practice you can even do it in your head for gigantic numbers with many digits! As a result, octal tends to be used as a kind of shorthand for binary. Many programming languages even have features that allow programmers to include octal numbers in their code.

So how do you convert octal numbers to and from binary?

The answer is that each octal digit always corresponds to exactly three binary digits in the following manner:

Octal	Binary
0	000
1	001
2	010
3	011
4	100
5	101
6	110
7	111

As an example, let us consider the octal number 6715:

- The 6 would correspond to the binary digits 110.
- The 7 would correspond to the binary digits 111.

- The 1 would correspond to the binary digits 001.
- The 5 would correspond to the binary digits 101.
- Putting the whole thing together, the octal number 6715 corresponds to the binary number 110111001101.

Likewise, if we wish to convert a binary number to octal we simply divide it into groups of three-digits starting from the righthand-side and locate the equivalent octal digit.

Consider the binary number 11011101110:

- The three rightmost digits in 11011101110 are 110, and this corresponds to the octal digit 6.
- The next three digits in 11011101110 are 101, and this corresponds to the octal digit 5.
- The next three digits in 11011101110 are 011, and this corresponds to the octal digit 3.
- The remaining digits in 11011101110 are 11, which we treat as 011, and this corresponds to the octal digit 3.
- Putting the whole thing together, the octal number 3356 corresponds to the binary number 11011101110.

Questions:

1. What is the denary equivalent of the octal number 361?

2. What is the denary equivalent of the octal number 506?

3. What is the denary equivalent of the octal number 7512?

4. What is the denary equivalent of the octal number 1074?

5. What is the denary equivalent of the octal number 531?

6. What is the octal equivalent of the denary number 527?

7. What is the octal equivalent of the denary number 908?

8. What is the octal equivalent of the denary number 431?

9. What is the octal equivalent of the denary number 7581?

10. What is the octal equivalent of the denary number 762?

11. What is the binary equivalent of the octal number 361?

12. What is the binary equivalent of the octal number 506?

13. What is the binary equivalent of the octal number 7512?

14. What is the binary equivalent of the octal number 1074?

15. What is the binary equivalent of the octal number 531?

16. What is the octal equivalent of the binary number 1001011?

17. What is the octal equivalent of the binary number 11010100111?

18. What is the octal equivalent of the binary number 1001010?

19. What is the octal equivalent of the binary number 10010101?

20. What is the octal equivalent of the binary number 11101100?

21. What is the octal equivalent of the binary number 111110101100011010001000?

22. What is the binary equivalent of the octal number 61702543?

Answers to Chapter 3 Questions

1. 241

2. 326

3. 3914

4. 572

5. 345

6. 1017

7. 1614

8. 657

9. 16635

10. 1372

11. 11110001

12. 101000110

13. 111101001010

14. 1000111100

15. 101011001

16. 113

17. 3247

18. 112

19. 225

20. 354

21. 76543210

22. 11000111100001010110011

Chapter 4: Hexadecimal

Just as denary is based on using powers of 10 and 10 different digits, binary is based on using powers of 2 and 2 different digits, and octal is based on using powers of 8 and 8 different digits, the hexadecimal number system (also known as base 16 or radix 16) is based on using powers of 16 and 16 different digits.

Given that we normally only have characters for only ten digits (0 to 9 inclusive), you might be wondering how we can get sixteen possible digits. The solution is to use the letters, A, B, C, D, E and F for the additional digits (Note: lowercase and uppercase letters are both commonly used, and the choice of case is **not** significant).

MOS KIM-1 Microcomputer Module - one of the very first affordable personal computers. Released in 1976, the KIM-1 sold for $245, and had just 1Kb of RAM and 2Kb of ROM. Note the hexadecimal display and keypad on the lower right of the circuit board:

The following table shows the digits used in hexadecimal numbers and their denary equivalents:

Hexadecimal Digit	Denary Equivalent
0	0
1	1
2	2
3	3
4	4
5	5
6	6
7	7
8	8
9	9
A	10
B	11
C	12
D	13
E	14
F	15

Consider the hexadecimal number 2EB:

65536s $(\times 16^4)$ $\times 65536$	4096s $(\times 16^3)$ $\times 4096$	256s $(\times 16^2)$ $\times 256$	Sixteens $(\times 16^1)$ $\times 16$	Units $(\times 16^0)$ $\times 1$
		2	E	B

- The rightmost digit (B), which is in the units column, represents 11 because B is the digit for 11 and $11 \times 16^0 = 11 \times 1 = 11$.
- The second from right digit (E), which is in the sixteens column, represents 224 because E is the digit for 14 and $14 \times 16^1 = 14 \times 16 = 224$.
- The third from right digit (2), which is in the two hundred and fifty sixes column, represents 512 because $2 \times 16^2 = 2 \times 256 = 512$.
- The total value of the number, **in denary**, can be calculated by added the column values together: 11 + 224 + 512 = 747.
- To sum up, the hexadecimal number 2EB represents the same number as the denary number of 747.

Consider the hexadecimal number C1A4:

65536s $(\times 16^4)$ $\times 65536$	4096s $(\times 16^3)$ $\times 4096$	256s $(\times 16^2)$ $\times 256$	Sixteens $(\times 16^1)$ $\times 16$	Units $(\times 16^0)$ $\times 1$
	C	1	A	4

- The rightmost digit (4), which is in the units column, represents 4 because $4 \times 16^0 = 4 \times 1 = 4$.
- The second from right digit (A), which is in the sixteens column, represents 160 because A is the digit for 10 and $10 \times 16^1 = 10 \times 16 = 160$.
- The third from right digit (1), which is in the two hundred and fifty sixes column, represents 256 because $1 \times 16^2 = 1 \times 256 = 256$.
- The fourth from right digit (C), which is in the four thousand and ninety sixes column, represents because C is the digit for 12 and $12 \times 16^3 = 12 \times 4096 = 49152$.
- The total value of the number, **in denary**, can be calculated by added the column values together: $4 + 160 + 256 + 49152 = 49572$.
- To sum up, the hexadecimal number C1A4 represents the same number as the denary number of 49572.

In the two examples that we have looked at so far, you will notice that the hexadecimal number and its denary equivalent both have about the same number of digits. In general, hexadecimal numbers tend to be the same length or slightly shorter than their denary equivalents.

In the rest of this chapter, we will look at how to convert numbers from hexadecimal to denary and vice-versa, as well as how to convert from hexadecimal to binary and vice-versa (you will shortly discover that converting between hexadecimal and binary is **very** easy - and it is for this reason that hexadecimal is used by programmers and computer scientists). Before doing so, here is a table of the powers of 16, which correspond to the values of the columns in hexadecimal numbers:

n	16^n
0	1
1	16
2	256
3	4096
4	65536
5	1048576
6	16777216
7	268435456
8	4294967296
9	68719476736
10	1099511627776
11	17592186044416
12	281474976710656
13	4503599627370496
14	72057594037927936
15	1152921504606846976

How to convert a hexadecimal number to denary

Here are the steps for converting a hexadecimal number to denary:

(1) Start with the rightmost digit of the hexadecimal number and multiply it by 16^0 (that is 1).

(2) Now move to the second from right digit of the hexadecimal number and multiply it by 16^1 (that is 16). Add this to the running total from your previous step to get a new running total.

(3) Now move to the third from right digit of the hexadecimal number and multiply it by 16^2 (that is 256). Add this to the running total from your previous step to get a new running total.

(4) Continue repeating this process until you have processed all the digits of the hexadecimal number, each time using the next higher power of 16. The total at the end is the denary value of the hexadecimal number.

Example: Converting the hexadecimal number 3B4D to denary.

(1) The rightmost digit is D which represents 13. So, $13 \times 16^0 = 13 \times 1 = 13$.

(2) The second from right digit is 4. So, $4 \times 16^1 = 4 \times 16 = 64$. The value from the previous step was 13, the running total is now 13 + 64 = 77.

(3) The third from right digit is B which represents 11. So, $11 \times 16^2 = 11 \times 256 = 2816$. The running total is now 77 + 2816 = 2893.

(4) The fourth from right digit is 3. So, $3 \times 16^3 = 3 \times 4096 = 12288$. The running total is now 2893 + 12288 = 15181.

(7) The hexadecimal number 3B4D corresponds to denary 15181.

How to convert a denary number to hexadecimal

Here are the steps for converting a denary number to hexadecimal (this method generates a hexadecimal number starting with the rightmost or lowest-value digit):

(1) Divide the denary number by 16, giving a whole number quotient and a remainder. The remainder is the first hexadecimal digit (do **not** forget to charge a remainder of 10 to 15 inclusive into the corresponding letter of A to F). The quotient is saved for the next step.

(2) Divide the quotient from the previous step by 16, to give a new whole number quotient and a remainder. The remainder is the next hexadecimal digit (do **not** forget to charge a remainder of 10 to 15 inclusive into the corresponding letter of A to F), and quotient is again saved for the next step.

(3) Continue repeating the process until you are left with a whole number quotient of 0.

Example: Converting the denary number 32351 to hexadecimal.

(1) We divide 32351 by 16. The quotient is 2021 and the remainder is 15. Hence the rightmost hexadecimal digit is F (which corresponds to 15).

(2) We divide 2021 by 16. The quotient is 126 and the remainder is 5. Hence the second from right hexadecimal digit is 5, giving a hexadecimal number, so far, of 5F.

(3) We divide 126 by 16. The quotient is 7 and the remainder is 14. Hence the third from right hexadecimal digit is E (which corresponds to 14), giving a hexadecimal number, so far of E5F.

(4) We divide 7 by 16. The quotient is 0 and the remainder is 7. Hence the fourth from right hexadecimal digit number is 7, giving a hexadecimal number of 7E5F. Since the quotient was 0, we know that we have finished the conversion process.

(5) Thus, the denary number of 32351 corresponds to the hexadecimal number of 7E5F.

Why programmers and computer scientists use hexadecimal

Hexadecimal numbers are used for the same reason that octal numbers are: hexadecimal numbers are short and easy-to-read, and can easily be converted to and from binary.

As with octal, once you have practiced, it becomes a simple matter to convert a hexadecimal number into binary or vice-versa. And, as is the case for octal, hexadecimal is therefore often used as a kind of shorthand for binary.

So how do you convert hexadecimal numbers to and from binary?

The answer is that each hexadecimal digit always corresponds to exactly four binary digits in the following manner:

Hexadecimal	Binary
0	0000
1	0001
2	0010
3	0011
4	0100
5	0101
6	0110
7	0111
8	1000
9	1001
A	1010
B	1011
C	1100
D	1101
E	1110
F	1111

As an example, let us consider the hexadecimal number 6BE5:

- The 6 would correspond to the binary digits 0110.
- The B would correspond to the binary digits 1011.
- The E would correspond to the binary digits 1110.
- The 5 would correspond to the binary digits 0101.
- Putting the whole thing together, the hexadecimal number 6BE5 corresponds to the binary number 0110101111100101 (or 110101111100101 if we do not wish to include the leading zero).

Likewise, if we wish to convert a binary number to hexadecimal we simply divide it into groups of four-digits starting from the righthand-side and locate the equivalent hexadecimal digit.

Consider the binary number 100011011101011:

- The three rightmost digits in 100011011101011 are 1011, and this corresponds to the hexadecimal digit B.
- The next three digits in 100011011101011 are 1110, and this corresponds to the hexadecimal digit E.
- The next three digits in 100011011101011 are 0110, and this corresponds to the hexadecimal digit 6.
- The remaining digits in 100011011101011 are 100, which we treat as 0100, and this corresponds to the hexadecimal digit 4.
- Putting the whole thing together, the hexadecimal number 46EB corresponds to the binary number 100011011101011.

Octal versus Hexadecimal

If you compare octal and hexadecimal, you can see that they both serve a similar purpose: they both offer a concise and easy-to-communicate shorthand for representing binary numbers.

An obvious question is whether one should prefer octal over hexadecimal or vice-versa?

- Many modern computing environments support both octal and hexadecimal numbering systems, so to some extent the choice is a matter of personal preference.
- Some tools or software may only support either octal or hexadecimal but **not** both. You may **not** always have a choice between octal and hexadecimal.
- For the reasons that are explained below, hexadecimal tends to be preferred on PC and Macintosh computers. Due to the ubiquity of these types of computers, you are probably more likely to encounter hexadecimal than octal.

Historically, octal tended to be mostly used on computers whose internal storage was organized around multiples of 3 binary digits ("bits"). For example, the PDP-8, ICL 1900, and IBM mainframes respectively used 12-bit, 24-bit, or 36-bit words, and an entire word on each of these machines could be displayed as a 4-, 8- or 12-digit octal number.

Likewise, hexadecimal tends to be most used on those computers whose internal storage is organised multiples of 4 binary digits ("bits"). PCs and Macintosh computers use storage units of 8-

bits ("bytes"), 16-bits, 32-bits, and 64-bits, and these binary numbers can be displayed as 2-, 4-, 8- or 16-digit hexadecimal numbers. As a result, hexadecimal tends to predominate on these platforms.

Viewing a file's contents using frhed, a "hexadecimal editor", on a computer running Microsoft Windows. Notice each 8-bit byte of data within the file is displayed as a two-digit hexadecimal number. The four-digit hexadecimal numbers on the left of the screen correspond to the offset relative to the start of the file:

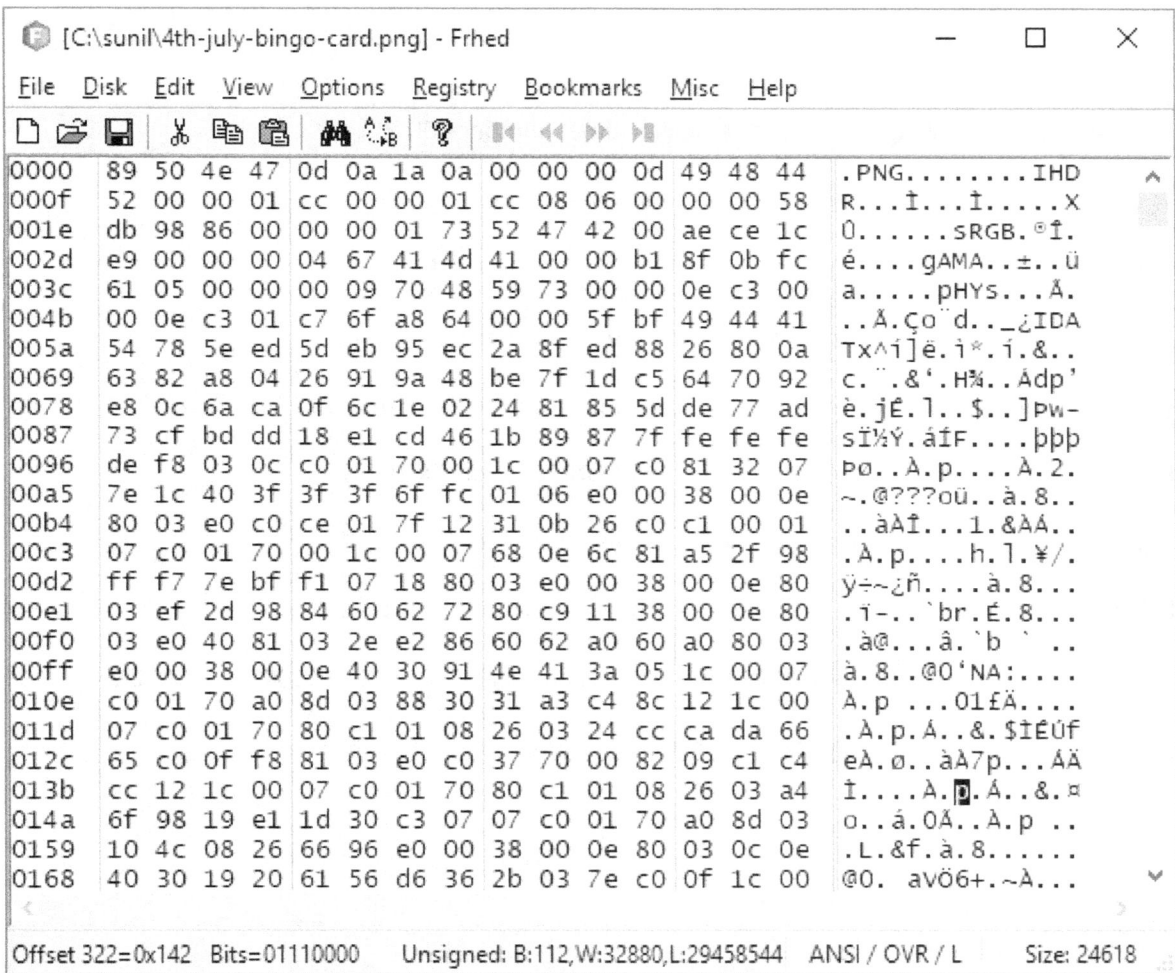

Questions:

1. What is the denary equivalent of the hexadecimal number 4A1?

2. What is the denary equivalent of the hexadecimal number 506?

3. What is the denary equivalent of the hexadecimal number 75E2?

4. What is the denary equivalent of the hexadecimal number F209?

5. What is the denary equivalent of the hexadecimal number BAC7?

6. What is the hexadecimal equivalent of the denary number 527?

7. What is the hexadecimal equivalent of the denary number 908?

8. What is the hexadecimal equivalent of the denary number 431?

9. What is the hexadecimal equivalent of the denary number 7581?

10. What is the hexadecimal equivalent of the denary number 762?

11. What is the binary equivalent of the hexadecimal number 5AB?

12. What is the binary equivalent of the hexadecimal number 3E6?

13. What is the binary equivalent of the hexadecimal number C0D2?

14. What is the binary equivalent of the hexadecimal number 142F?

15. What is the binary equivalent of the hexadecimal number 5F1?

16. What is the hexadecimal equivalent of the binary number 1001011?

17. What is the hexadecimal equivalent of the binary number 11010100111?

18. What is the hexadecimal equivalent of the binary number 1001010?

19. What is the hexadecimal equivalent of the binary number 10010101?

20. What is the hexadecimal equivalent of the binary number 11101100?

21. What is the hexadecimal equivalent of the binary number 111110101100011010001000?

22. What is the binary equivalent of the hexadecimal number 8A702C4E?

Answers to Chapter 4 Questions

1. 1185

2. 1286

3. 30178

4. 61961

5. 47815

6. 20F

7. 38C

8. 1AF

9. 1D9D

10. 2FA

11. 10110101011

12. 1111100110

13. 1100000011010010

14. 0001010000101111

15. 10111110001

16. 4B

17. 6A7

18. 4A

19. 95

20. EC

21. FAC688

22. 1000101001110000010110001001110

Conclusion

Well done for getting to the end. I hope you enjoyed this book!

For more binary, octal, hexadecimal, and related information, please go to: http://www.suniltanna.com/binary

You might also be interested in my book on more advanced binary topics: Advanced Binary for Programming & Computer Science: Logical, bitwise and arithmetic operations, and data encoding and representation.

If you enjoyed this book or it helped you, please post a positive review on Amazon!

To find out about other educational books that I have written, please go to:

- For computing books: http://www.suniltanna.com/computing
- For math books: http://www.suniltanna.com/math
- For science books: http://www.suniltanna.com/science

And remember: If you enjoyed this book or it helped you, please post a positive review on Amazon!

www.ingramcontent.com/pod-product-compliance
Lightning Source LLC
Chambersburg PA
CBHW081649220526
45468CB00009B/2599

9781722300548